1 2 3 4 5 6 ... 22 37 49 50

50 QUICK & EASY WEBSITE TIPS

by Teena Hughes

© 2014

HTTP://EASYCONTENTMARKETING.COM

50 Quick & Easy Website Tips:
Simple Tips to Make Your Website Irresistible
Pepper The Internet,
Be found everywhere

ISBN-10: 0-9943971-3-5 • ISBN-13: 978-0-9943971-3-3

Author / Publisher
Teena Hughes
42MonkeysPublishing.com
PO Box 424, Burleigh Heads QLD 4220
Australia

Cover design by Teena Hughes

For Information CONTACT Teena Hughes in Australia (+61) 408-801-808

Visit TEENA on the web
http://42MonkeysPublishing.com

Disclaimer
Any content and information within this book is for informational and educational purposes only. The internet is changing faster than the speed of light, and some information in this book may be out of date by the time you read it. Please do your research before embarking on major changes to your business, and surround yourself with experts on the topics you need to become familiar with.

Dedication

This little book is for website owners, business owners, and everyone who needs to have an internet presence.

If you've been tearing your hair out trying to work out what to do now you've got a website, and what all that gobbledegook means, this simple series of tips will help.

Each tip explains one simple thing to help your site - and your business.

Ready? Let's get started!

cheers,
Teena!

1 2 3 4 5 6 ... 22 37 49 50

The internet changes as quickly as a chameleon — new rules, new tools, new amazing technologies, so I thought it would be a great idea to update this book whenever new information is available.

To make sure you don't miss out, please visit:

Want Free Lifetime Updates?

http://teenahughesonline.com/50tipsupdates

50 QUICK & EASY WEBSITE TIPS

SIMPLE TIPS TO MAKE YOUR WEBSITE IRRESISTIBLE!

BY
TEENA HUGHES

1 2 3 4 5 6 ... 22 37 49 50

Do you need TRAINING on how to do these tasks?

http://teenahughesonline.com/50TipsCourse

1 2 3 4 5 6 ... 22 37 49 50

Ready?

Turn the page for
Tip #01

It's extremely important to have a way to KEEP IN TOUCH WITH folks who visited your site while you were sleeping (or just busy).

An easy way to do this is to add a SIGN-UP BOX on each page to offer web visitors more information.

The visitor provides their email name, you provide more info about your website, products and services.

1
Sign-Up Box

Installing web statistic software a.s.a.p. is a great way to find out WHERE your web traffic is coming from (keyword searches, other websites, advertising campaigns etc). Try Google Analytics and Google Webmaster Tools.

Install them sooner rather than later in order to capture lots of stats. For REALLY simple stats, install statcounter.com plugin on your Wordpress site.

2

Web Stats

Offering a bonus of great perceived value in exchange for someone's email name is a great way to show them you know what you're talking about. Make it something the visitor really wants or needs to help solve an immediate problem they may have.

3

Sign-Up Bonus

Make a list of 5-10 solutions to help your web visitors' problems. Turn each one into a helpful email, then using Newsletter software (like Aweber) turn those into an auto-responder series.

4 Auto-responders

Once someone signs up, they can be drip-fed the email series for days, weeks or months, which helps them get to know you, your products and services.

Consider sending out a weekly, fortnightly or monthly newsletter to folks who've signed up on your site.
Be informative and entertaining, share Case Studies where you've helped others, Testimonials, and plenty of tips.
This helps show your expertise and knowledge and may earn trust.

5
News!

Sign up for several Social Media accounts such as Twitter, Facebook Business Page, Pinterest, Linkedin, Google + etc.

Signing up and trying to be active on too many accounts will create overwhelm in the blink of an eye, so I suggest starting with one until you get the hang of it.

When you're comfortable with one, consider adding another.

6
Social Media Accounts

Each page and blog post needs to have visible linked icons for:

* Twitter
* Facebook
* Google +
* LinkedIn
* Youtube
* Pinterest etc.

This helps website visitors to click to go read what you've written on those other websites, to complement your site's information.

7 Social Media Icon Links

Add functionality to your website which allows visitors to "like", "tweet" and share information they find particularly helpful.

A simple way to do this is to add a plugin / widget (if you're using Wordpress). These should appear on every page and blog post.

8
Social Follows

Web pages are for standard information about a business or website. Blog posts are a great way to keep adding interesting information & tips, keeping the site active which keeps the search engine robots coming to investigate. The more often you write blog posts, the more often your site will be visited by the many search engine robots (SEO).

9
Add Regular Blog Posts

Yes you do need to write blog posts to boost your website for SEO (Search Engine Optimization - being found on the internet).
Don't roll your eyes — it's easier than you can believe!

Write a list of questions you've been asked by potential clients.
Write one blog post for each question, and answer it as though you were talking to a friend. Do this WEEKLY (at least).

10 Easy Blog Posts

Imagine someone lands on your website from something they found online, they like what they read, and want to contact you immediately. Do they have to go click the Contact link in the Menu (or worse still — SEARCH for that), or can they find your phone number and contact details at the top or bottom of every page, or in the sidebar? They can? Fantastic! Bravo!

11
Contact Details

Make sure your About page has a photo of you and the staff (if you have any), and remember to make it interesting and memorable.

There are plenty of 'dry' About pages online which make your eyes glaze over. Don't be one of those. Please. I mean it.

Share an unusual fact for each person showing their 'human' side and sense of humour.

12

The About Page

Visitors land on your website pages after reading about you online, offline, URL (link) on a taxi, offline advertising etc.

EVERY PAGE and post needs to be seen as the "home" page (or a "landing page") by clearly showing the business or website NAME, contact details and a way to sign-up.

Each page or post should also have ONE THEME or TOPIC, and focus on that.

13

Every Page is a Landing Page

Search engines love it if you leave little 'breadcrumbs' for them to follow, by linking a series of words on one page to a different page on your site. This is called 'hyperlinking' and if you select a keyword phrase and link it to another page ABOUT those keywords, you get extra points :-)

14 Hyper-linking

Please please please don't put "Last updated on …" on your website. Seriously. I mean it. If you forget to update your site or just get way too busy, no-one wants to see a date from 2005. I mean it. There's really no reason to put "last updated" on the majority of website pages. Blog posts show their "posted on" dates, and that's ok :-)

15 Last Updated

Whether you have a shopping cart or use Paypal buttons, make sure to test them. Set up a sample product for $1.00, now go through the entire purchase process - check each step in the process, read the messages in the sales process online, and double-check the automated emails.

Imagine you're a first time visitor and customer on your site — were YOU happy?

16

Selling something?

If someone buys from your site, take the time to send a friendly email and ask them about their experience.
Were they happy?
Were there any problems?
Could they make any recommendations to improve the sales process?
Customers love to know their opinion is valued, so remember to ask for it.

17
Ask For Feedback

When writing on your website, imagine that you're talking to a friend who doesn't really know what you do.
Keep the language simple and easy to understand, and not too technical, and this will make it easy for all visitors - whether they have some or no knowledge of your expertise.
If in doubt, make it simpler.

18

Use Plain English

Video is HUGE for business and websites, whether as Testimonials, product and service explanation or introduction videos, staff member video introductions, video thank you to potential clients you met - the list goes on. By creating a Youtube Channel for your videos, you can be found on the second largest search engine there is — Youtube. Don't miss this step. It's a biggie.

19

Create Youtube Channel

Got a smartphone or computer with camera? Consider making video testimonials for products and services YOU use in your own business. Post on your Youtube Channel then send a link to the company in question.

20

Make Video Testimonials

Do this as a warm and generous Testimonial — and if they offer a link back to your site, that's a huge added bonus.

Each time you post a video, do a bit of video SEO to help it be found amongst the millions of other videos online.

Use the Description box on Youtube — post a link back to a specific page/post on your website where this one video appears, write a great long description and use your keyword phrases too. Include all the ways people can find you - website & all social media accounts.

21

Video SEO

When you have a few videos on Youtube, create Playlists where you list all the videos with a similar theme, for example Testimonials, How To's, Tips, Newsletters, Sales Videos etc.

As you create new videos, add them to the relevant playlists and don't forget to send messages to your Youtube subscribers to let them know.

22

Video Playlist

People online are searching for answers to their problems, and your potential clients and customers are no exception.

Back in #4 I suggested writing 5-10+ questions you've been asked.

Now it's time to turn those questions and answers into videos - one question per video.

Post on Youtube, your site, and other video portals so more people will find you.

23

Video Q&A

Create a video which is an introduction to your website, walk the visitors through the sections of your site, and show how easy it is to contact you. The video only needs to be 2-3 minutes long, not War And Peace, and can appear on the Home page and in the sidebar on other pages and posts. Make it as easy as possible for your potential customer to find out more about you.

24
Website Video Intro

If you create a Newsletter (or even if you don't) turn it into a video and post on Youtube and your site (& other video portals). Include something interesting (solve problems), something fun, and be casual and relaxed. Ask the viewers/readers to send you questions via Youtube or your site or social media accounts. Then create a video based on each question and answer - this is a real winner!

25

Video News

Come up with an idea for 60 second Tips on your area of expertise, which could help your potential customers.

Using your smartphone (or computer) create quick videos while you're out and about, which shows you're a real person going about their work day.

The tips can also be posted on social sites and Pinterest, to increase your potential clients further.

26
Video
Tips

Add a video Site-map to your website (search how to do this, or use a plugin for your Wordpress site). A video site-map is a great treat for the search engine robots who'll come visiting your site. This will help them to 'index' your videos.

Make sure to post the videos on your social sites, video portals, turn them into Playlists, and even burn them onto DVDs to give to potential clients.

27
Video Site-map

When you have about 10 video Tips (or other series of videos), consider turning them into a DVD.
These can be handed out at events, given away on your site (for just the cost of postage), sent to clients as a bonus for their loyalty, etc.
Create your own or use a DVD distribution & fulfilment service to do the posting for you. Reuse your content over & over & over.

28
Video
DVD

If you're new to websites and the internet, you might be tempted to copy photos from search engine searches. DON'T. I REPEAT - DON'T.

This is illegal and one misuse of someone else's photo could cost you many thousands of dollars.

Use photos for which you have a licence (you may need to pay), or your own photos. But DON'T STEAL SOMEONE ELSE'S.

29

Using Images on a Website

Before uploading images to your website, rename them to match the web page or blog post name. Use hyphens (never underscores, never spaces) for the jpg name.

If your post is named great-aussie-cafe-in-new-york.html name the jpg to match it - great-aussie-cafe-in-new-york.jpg

30

Naming Images FIRST

Images downloaded from your smart phone or camera can be HUGE, so before adding them to a website or blog it's a great idea to reduce the WEIGHT. Do this by opening in image editing software and "save for web". This reduces the weight without affecting the quality. Images on websites don't need to be as high quality as a glossy fashion magazine (20-100kb each is good).

31
Image Size

Whilst adding an image to a page or blog post, remember to type some descriptive words in the ALT tag, for example: "Photo of great Aussie cafe in New York".
If some folks turn images "off" they'll see the Alt tag. It will also be visible when a mouse is hovered over it on the screen.
This helps everyone viewing a web page or post.

32
Image Alt Tag

Don't fall into the trap of only putting a photo on a page or post, with no supporting text.

At this point, search engines cannot read text on images, so it's very important to write on the page or post to describe the photo or image.

The more text, the better the chance the page or post can be catalogued by its content.

This is a good thing to do - VERY good :-)

33

Use Text Plus Image

If your neighbour's adult son's wife's cousin's teenager says they "know SEO", wait a minute.

If your hairdresser's brother's Swedish girlfriend's gay friend and partner says THEY "know SEO", hang on a tic.

Before agreeing to anyone touching your website for SEO purposes, get REFERENCES from businesses. Use my "14 Questions to Ask an SEO Guru" Checklist, just to be on the safe side. Promise? Cross your heart? Good. Phew!

34

SEO Guru

35 Repurpose Everything

It's crucial to repurpose everything you create: most people write a blog post, or an article, or even a PDF about their business, products or services. They generally publish it in ONE place, but they don't realise they're missing out on a HUGE opportunity!

There are millions of people looking at millions of sites, looking for what you've got. Get it out there - help them to find YOU.

How to **repurpose everything you create?**
Turn a blog post into a PDF, a slideshow, a checklist, a video, an audio or podcast — and distribute them all far and wide across the internet.

There are many, many portals for each type of content, so don't post one thing once, post each thing to 5, 10, or 50 different locations online.

THIS is how you get known online - it takes courage and perseverance, are you ready? Yes? Woohoo!!

*Pepper The Internet - a phrase I started & introduced many years ago!

36
*Pepper The Internet!

If you want your website and business to succeed, a very simple step is to create a Calendar or two.

A yearly blog and/or marketing calendar might be too daunting, so start with a monthly one, just make a simple plan to post one new thing once a week, then increase it if it gets too easy. Having a calendar CAN help you organise your website activity (and other marketing), and consistency is the KEY. We've all heard it, now here's your chance to test it out.

37
Website Calendar

To make sure you can TELL there have been improvements in your website traffic, leads (sign-ups, Likes etc) and sales, you need to TRACK & MEASURE them.
This could be as simple as checking your web stats to see where traffic is coming from, or noticing an increase in sign-ups.
Create a list or a spreadsheet and note dates & changes - this will encourage you to keep making changes.

38

Track & Measure

If you're really into statistics and tracking, you may also like to set up Google Analytics to track and measure the performance of campaigns.

If Google Analytics seems complex, hire a specialist to create the campaigns for you, and to set up Reports to be emailed to you.

Using another's expertise in THEIR field is a great way to build your business.

39

Google Analytics Tracking

If there are things on this list you'd love to try but you just don't have the time, skill or patience, spend a small amount of money to get someone else to to the tasks for you. Never has it been easier to hire Virtual Assistants in your own or other countries, to help you grow your business. Let them streamline your processes and make your job easier every day.

40
Get Help! Outsource

Many of the suggestions in this book can be automated. There may be services already available to do a lot of these tasks and processes, so do a bit of research and find something which works best for you.

41

Use Automation

No need to do ALL the tips at once — start with one, find a solution; when it's running smoothly try another and automate it. You'll be so glad you did :-)

I use a password management system online which I can login to from anywhere in the world.

It's robust and secure, and highly regarded by huge corporations.

This tip has saved me constantly struggling to come up with new and clever passwords for the hundreds of systems I use.

It's saved my bacon once or twice, it might just save yours too (it's listed in the Resources).

42

Password Management

It's possible to automate some (but not ALL) of your social media, and there are excellent tools to help you do this.

This can save a bucket-load of time and keep your "social" stuff working while you sleep.

For example, come up with 10, 20, 30 short tips for all potential customers. Turn them into Tweets, and schedule them for a month in advance.

Easy peasy!

43
Automate Social Media

I think we're all in the same boat — we can become slaves to email, checking it umpteen times a day.
Try setting two time slots to check email daily.
Don't read any email for the first two hours of your day — you will be super productive!
Try 10:00 - 10:30 and 3:00 to 3:30 while you have a tea or coffee break. Then QUIT out of the software. I mean it. Try it. You just might like the extra energy and focus you have :-)

44

Use Email LESS

45

Dream "To Do" List

Last thing at night, write down your "dream" achievements for the next day (or do this first thing in the next morning).
What would you absolutely LOVE to get done today? How will that make you feel?
Avoid writing long checklists of items - that will overwhelm you before you start.
Do the DREAM task/s FIRST to make you feel fabulous and a great ACHIEVER!

It might sound a bit ooroo-guru or flaky, but this is definitely worth trying. Write down the most positive things you'd like to be able to describe yourself as, for example:

* I am successful every day.
* I am patient with others.
* I am improving myself constantly.
* I achieve all my goals quickly and easily.
* I am a strong, happy, financially independent woman/man.

46 Positive "I Am" Statements

Say these out loud or in your mind every morning when you're fresh and vibrant. **BELIEVE IN YOURSELF!** I know I believe in you, and so do others.

Got tasks in your head chasing each other at breakneck speed, but you're toooo busy to get to them?

Each Monday at 9:00am, PICK ONE TASK that's been bugging you. Write it, print it, stick it up on the wall as your FOCUS for THIS WEEK.

How long will it take, do you think?

47

Pick One, Just One

How long per day? How much can you get done in 5 days? Good. Now you know. Start it straight away. When it's done, PAT YOURSELF ON THE BACK! Rinse and repeat next Monday with a NEW task.

Whether you have a large business or a small one-person business, it's extremely important to list your daily & weekly tasks. This is your "insurance" for when you need to take a holiday or if you get sick and need to get someone to help out.

48

List Your Tasks

If your tasks are listed, it will make it easier to hand over the reins whenever you need to … after you've finished Tip 49!

For every task performed in every business, there should be a Process documented, written down, or recorded on video (SOP = Standard Operating Procedure). Don't have anything like this? Think you don't need it because you're a one-person business?

Wrong. Everyone needs this. With Processes in place you can HAND OVER TASKS to others to follow to a "T". Free your brain up for the things YOU'RE great at. Got 10 minutes? Create your first Process document, diagram or video. Bravo!

49
Create Processes, SOPs

There are thousands of tips out there to help people in all walks of life, for all types of businesses and adventures.
The problem I see all the time is OVERWHELM.
The way to fool this problem is to FOCUS ON ONE THING at a time. Don't let the others crowd your brain. Do one thing, do it well.
Move on the next, do it well too.

The way forward for everyone is simply **ONE STEP AT A TIME**. I wish you all the very best of success with whatever is in your future!

50

The Last Tip!

1 2 3 4 5 6 ... 22 37 49 50

If you got to this page, I'm thrilled to bits! If you try only one thing I've mentioned, you'll be one step further ahead than before you started. If you take just one step forward each day, you'll be WAY ahead of your competitors in no time. How will that make you feel? I have a surprise bonus — two final tips for you ...

Thank you for reading this far! But wait! There's more!

The 51st step in this series of website adventures is to TAKE ACTION.

No more procrastinating.

No more, "I've got no time!"

No more, "Too much to do!"

No more, "I'll never be able to do that!"

The time is NOW and to be found online by potential clients and customers, YOU need to take ACTION.

Ready?

Fantastic!! Woohoo!

Start TODAY!

51

Take Action!

When you launch your website, go through it one page at a time and CLICK EVERY LINK. Make sure the links go where they're supposed to, because your website visitors may not have enough patience to write and let you know there's a broken link.
If ANY link goes to a different website, it MUST MUST MUST open in a new browser window. Trust me, this is very important.

52
Check All Site Links

My goal was to simplify a few concepts and provide some confidence for you to feel like you can take a step forward with your website or business.

If any one of these pages helped, I'm thrilled to bits - thanks so much for reading to the end!

If you'd like access to the list of Resources, please pop over to the next page.

I hope you've enjoyed my book

1 2 3 4 5 6 ... 22 37 49 50

Would you be so kind to ...?

If you have enjoyed my book, I'd be ever so thrilled if you'd post a Review on Amazon — I'd be ever so grateful!

If you're a clever clog and would like to make a video or email me with a Testimonial to tell me you've enjoyed my tips, that would be fantastic too!!

Thanks you so much! Cheerio for now!
from Teena Hughes

1 2 3 4 5 6 ... 22 37 49 50

I'm updating my list of Resources all the time with the latest and greatest, so rather than print them and have them be out-of-date, I've put them on a webpage for you to visit.

My List of Resources

Sign up as a purchaser of my Kindle book to receive Private Access in the first email I will send to you.

1 2 3 4 5 6 ... 22 37 49 50

Want

Free Lifetime Updates?

The internet changes as quickly as a chameleon — new rules, new tools, new amazing technologies, so I thought it would be a great idea to update this book whenever new information is available.

To make sure you don't miss out, please visit:
http://teenahughesonline.com/50tipsupdates

1 2 3 4 5 6 ... 22 37 49 50

The Author

1 2 3 4 5 6 ... 22 37 49 50

Hi, my name is Teena Hughes and I've been helping people launch their businesses and websites online for many years. The first time I saw a computer mouse was when I was hired by an advertising agency in Paris to work on Microsoft Word Version 1 — in French — on an old computer (way before Windows; ever heard of DOS?). I had to learn French so I could translate the Manual, which would have been so much easier if the internet was around :-)

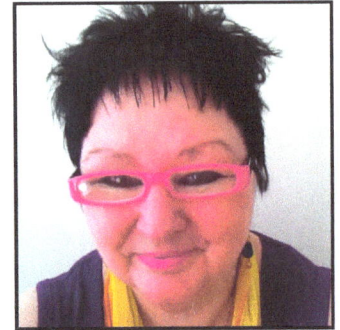

About Teena Hughes

1 2 3 4 5 6 ... 22 37 49 50

I never mastered the mouse that first day; I plugged it in, and when I ran out of room on the desk, I ran the mouse down the leg of the table. True story! I know you're laughing at me — and I'm laughing too!

I then wrote tips and notes to help the next person who might do my job, and so began my desire to create computer instructions, lessons and courses to help other people (before they started tearing out their hair in frustration).

1 2 3 4 5 6 ... 22 37 49 50

With the help of the internet I can now publish books to help even more people, which is truly fantastic!
If you've signed up already at

http://teenahughesonline.com/50tipsupdates

you'll be on my early-bird list to also get updates to this book, as well as advance notice when I have new books and courses.

I look forward to chatting with you soon! Whatever you do, remember to smile, laugh and enjoy every day. A smile is infectious, and it's always great to share the love!

1 2 3 4 5 6 ... 22 37 49 50

Do you need TRAINING on how to do these tasks?

Do you wish you could follow step-by-step instructions, and watch easy-to-follow video training? My Course might be just what you're looking for. Read more here:

http://teenahughesonline.com/50TipsCourse

I look forward to chatting with you soon!
Whatever you do, remember to smile, laugh and enjoy every day.
A smile is infectious, and it's always great to share the love!

1 2 3 4 5 6 ... 22 37 49 50

Please remember to do your best AND be your best every day, that's your #1 job!
And don't forget to enjoy your journey!

The End ... for now ...

www.ingramcontent.com/pod-product-compliance
Lightning Source LLC
Chambersburg PA
CBHW052053190326
41519CB00002BA/202